AF538333
Greenland
Norwegian Sea
Finland
Europe
Atlantic
Ocean
Senegal
Africa

Carried on the Wind
It Starts with a Seed

Sheri Mabry

illustrated by
Kristina Jones

ALBERT WHITMAN & COMPANY
CHICAGO, ILLINOIS

Somewhere,
in a tiny yard,
in a small town,
in the big world,
someone sighed.

Dandelions provide nectar for up to 100 insect species. And some animals—rabbits, ground squirrels, and mice—eat the seeds, leaves, and roots.

**Like a slight breeze,
the sigh blew the lightweight seeds
into twirling and spinning and floating.**

Dandelion seeds can travel over 100 miles on the wind. The fluff acts like a parachute. When a breeze—or a sigh—blows the seeds, they scatter through the air.

These dandelions are in northern Alaska.

A hummingbird saw the white puffs, swooped down, and scooped some up. She flew the fluff to her nest and wove it in so she could rest and lay eggs on top of it.

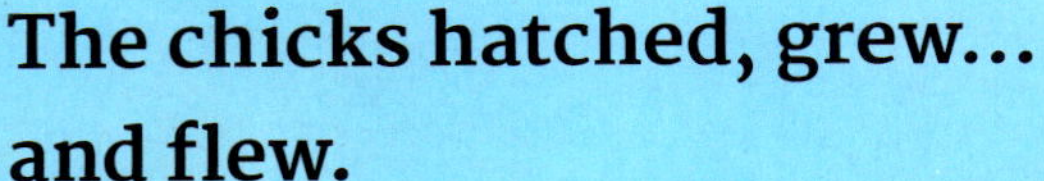

**The chicks hatched, grew...
and flew.**

When the baby hummingbirds are grown, some migrate. This small Rufous hummingbird flew alone from Alaska down the coast of the Pacific Ocean and across the Rocky Mountains to Central America, maybe Panama.

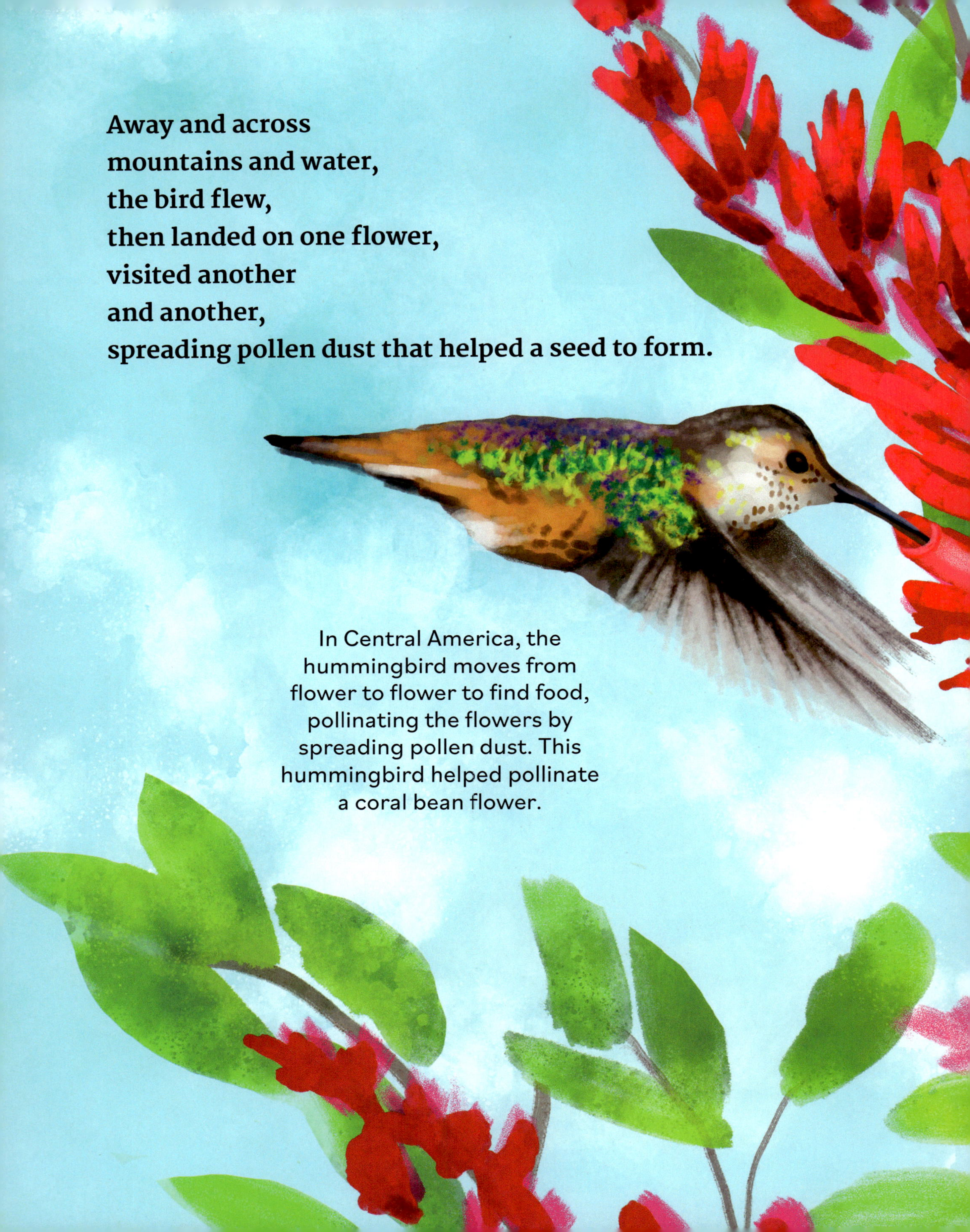

Away and across
mountains and water,
the bird flew,
then landed on one flower,
visited another
and another,
spreading pollen dust that helped a seed to form.

In Central America, the hummingbird moves from flower to flower to find food, pollinating the flowers by spreading pollen dust. This hummingbird helped pollinate a coral bean flower.

Rain fell on the seeds.
One seed rolled into a stream,
which flowed into a river,
which rushed into the sea.

The seed floated a long way on swelling waves, through sunsets and moonrises.

When a plant is pollinated, it makes seeds. Seeds from the coral bean flower scatter and fall. Sometimes, they fall into rivers, lakes, or oceans. Drift seeds can float in salt water for at least one year, over a long distance. Their thick coats protect them and allow them to float. This seed drifted across the Caribbean Sea and the Atlantic Ocean towards West Africa, maybe Senegal.

The seed settled onto a sandy beach.

Bugs started to nibble,

When drift seeds end up on sand,
sometimes young insects called
doodlebugs find them. Adult
doodlebugs are called antlions.

then grew,
sprouted wings,
and flew.

Birds ate the bugs.
When the birds were full,
they sipped and dipped
in water.

Some birds, like these
hoopoes, eat antlions.

**Not far away,
an elephant heard
the splashing bugs.
Her big ears flapped.
She sniffed.**

African savannah elephants have excellent hearing. They also have millions of scent receptors that help them smell water far away—even water buried underground.

The elephant let out a low rumble
that vibrated for miles
through the earth
and was felt by padded, soft feet.

Elephants send messages to each other by making vibrations in the earth that other elephants feel through the soft skin on the pads on the undersides of their feet or by laying their trunks on the ground. Elephants may exchange these vibrating messages even when they are 20 miles apart.

The herd heard its leader's rumble.
It followed her—
slowly, steadily—
to the water.

An elephant herd is led and protected by a matriarch, often the oldest female.

All together, they dug.
The hole filled with more water
from deep below.
The elephants drank and showered and sprayed.
Droplets landed on dirt,
turned into mud.

**Butterflies flew to the drops.
They licked
until they were ready to fly.**

In Africa, painted lady butterflies, like butterflies everywhere, need water to survive. They lick moisture through their tubelike tongues.

When the light was right,
the butterflies fluttered into the wind—
a blanket of soft speckles
winging away and across the sky.
Across mountains and water.

Less daylight and food help signal to butterflies to migrate. Painted ladies may fly as far as 9,000 miles round trip.

The butterflies landed
in a field of flowers,
in a tiny yard,
in a small town,
in the big world...
where someone,
now a bit older
and a little taller,
rested against a tree.
A butterfly lit upon a knee.

And someone...
smiled.

Butterflies sometimes land on people, often looking for salt on their skin, which they need to survive. They only land if they feel it is a safe place.

Author's Note

This story reminds us about our connection to nature and our part in it. We can support nature, even with something as simple as growing dandelions that produce nectar in our yards. Each of us impacts our world through our presence in it. The child in this story simply sighed, and that breath began the series of small events that make up this book.

The events in this book probably wouldn't happen like this in real life, if they happened at all. The journey described and the path it takes aren't real. The butterflies in this story left Senegal, spent time in Finland, flew across the Norwegian Sea, and crossed part of Greenland into the Arctic Circle and then to Alaska, where the story began. The butterflies that landed in Alaska are not the same ones that started the journey in Africa. They are a different generation.

It is important to remember that we are all connected to nature and that everything we do causes something else to happen. Our thoughtful connection to nature helps make the world a more beautiful place.

Chaos Theory

Have you ever noticed how one tiny action can set off other actions? It's like when you throw a pebble into a pond and ripples spread out from where it falls in the water.

There's a name for this. Scientists call it chaos theory. This is how one tiny event can cause a ripple effect. As you notice in this story, the tiniest action—the child's sigh—sets off a series of events. That sigh affected nature all around the world.

Chaos theory helps us realize how even tiny actions, like the child's sigh, make a big difference in the world. It reminds us how we are each connected to our world.

Glossary

matriarch: In the elephant's world, a matriarch is the leader of the family. She is usually the biggest female and the oldest elephant of the group. Her job is to guide the other elephants, including the young ones. She teaches them how to stay safe and find food.

migration: Moving from one place to another, usually to find food or a safe new home. Animals and insects that migrate often travel long distances.

nectar: A sweet liquid found in flowers, such as dandelions, that butterflies and other insects sip.

pollen: A powdery material that comes from flowers, trees, and plants. It helps plants reproduce.

pollination: When animals or the wind carry pollen grains between the same kinds of plants and the pollen makes seeds for new plants to grow.

salt water: Water that contains more than 3 percent salt, such as the water in our oceans. More than 97 percent of our water on Earth is salt water that animals and people cannot drink.

scent receptors: Structures inside an elephant's nose that help it recognize smells. Elephants have a very strong sense of smell. Their trunks pick up many different scents, even from far away.

For: Whitnie-whose first breath changed my life profoundly and beautifully. You are a blessing to the world. I'm so grateful for you. Love, Mom.—SM

To my daughter, who reminds me that our actions shape the world we inherit—may you always cherish and protect the beauty of nature.—KJ

Library of Congress Cataloging-in-Publication data is on file with the publisher.

Illustrations by Kristina Jones
First published in the United States of America in 2024 by Albert Whitman & Company
ISBN 978-0-8075-7372-3 (hardcover)
ISBN 978-0-8075-7373-0 (ebook)

Printed in China
10 9 8 7 6 5 4 3 2 1 WKT 28 27 26 25 24

Design by Rick DeMonico and Shane Tolentino

For more information about Albert Whitman & Company, visit our website at www.albertwhitman.com

Canada
Alaska
Rocky Mountains
Pacific Ocean
United States
N
W
E
S
Central America
Caribbean Sea
Panama